CUTTING-EDGE CAREERS™

CAREERS IN
ROBOTICS

Paul Kupperberg

ROSEN
PUBLISHING®

New York

With appreciation to my editor, John. I never know what he's going to throw my way next, but I do know it will always be interesting.

Published in 2007 by The Rosen Publishing Group, Inc.
29 East 21st Street, New York, NY 10010

Library of Congress Cataloging-in-Publication Data

Kupperberg, Paul.
Careers in robotics / Paul Kupperberg. — 1st ed.
    p. cm. — (Cutting-edge careers)
Includes bibliographical references and index.
ISBN-13: 978-1-4042-0956-5 (library binding)
ISBN-10: 1-4042-0956-5 (library binding)
1. Robotics — Vocational guidance. I. Title.

TJ211.25.K87 2007
629.8'92023 — dc22

                                                    2006020189

Manufactured in the United States of America

On the cover: A robot plays soccer in Germany during the 2006 "RoboCup" project which pits a team of robot soccer players against human competitors.

CONTENTS

INTRODUCTION

[I]n late April 2006, students from all around the world gathered at the Georgia Dome in Atlanta, Georgia, for a competition very different from the athletic contests that the stadium usually hosts. This was a contest not about strength, speed, or stamina, but about smarts: the FIRST Robotics Competition.

The FIRST Robotics Competition was created by inventor Dean Kamen 16 years ago. Kamen is the creator of the Segway Human Transporter and numerous medical and scientific inventions. His goal in organizing the competition was "to inspire an appreciation of science and technology in young people, their schools and communities," as quoted in FIRST promotional materials. When it began in 1992, 28 teams meeting in a local New Hampshire school gym participated. The 2006 contest attracted some 23,000 students in 1,125 teams to 33 regional events, leading to the championship event in Georgia.

While one of the better known, the FIRST Robotics Competition is only one of many such competitions in the United States and around the world that are sponsored by private industry, the military, and the National Aeronautics and Space Administration (NASA). They are all designed to spur interest in the field of robotics.

Why the emphasis on robots over other areas of technology? Because, while robotics has its roots in science fiction, popular science,

A contestant in the 2006 FIRST Robotics Competition held in Atlanta, Georgia, readies his entry. Some 1,125 teams competed for scholarship prizes in this annual event.

and the fevered imaginations of futurists, the last three decades of the twentieth century have seen the "mechanical man" graduate from a flight of fancy in comic books, dime-store paperbacks, and B-films to a practical, hardworking tool of everyday life.

A robot is a mechanical device that can perform—either under human guidance or through the direction of a predefined program or set of guidelines—specific physical tasks. This is accomplished through the science of robotics, which combines software, mechanical manipulators, sensors, controllers, and computers to allow for programmable automation of tasks.

The idea of robots dates back as far as c. 2500 BC in Egypt and has continued to fascinate mankind ever since. Egyptian priests invented the concept of a "thinking machine" when they placed men inside statues that served as oracles—mouthpieces for the wisdom of the gods. Though this was a bit of fakery, it nevertheless spawned the idea of inanimate objects being provided with the powers of speech and intelligence. Their first appearance in literature seems to have been in Homer's eighth-century BC epic poem *The Iliad*. It mentions an *automaton*—a mechanical device, usually powered by water, wind, or clockworks—and a *simulacra*, or a device created in the image of a living thing. The first practical application of robotics was most likely the 270 BC invention by the Greek inventor and physicist Ctesibus of a clock that measured time as a result of the force of water falling through it at a constant rate.

Over the centuries, automatae continued to fascinate people the world over. Crude clockwork automatae were created, including an entire mechanical orchestra in third-century BC China and, reportedly, devices built by artist and inventor Leonardo da Vinci (1452–1519) in AD 1495. In 1822, British scientist Charles Babbage (1791–1871) invented his difference engine, a device capable of performing mathematical calculations. For this achievement, he would one day be honored with the title "Father of the Computer."

Robby the Robot, costar of the 1956 science fiction epic film *Forbidden Planet*, was created in the MGM Studios prop shop. Though only a costume worn by an actor, Robby helped popularize the idea of friendly, intelligent robots in the public imagination.

The term "robot" was coined in 1921 by Czech playwright Karel Capek (1890–1938) in his play *R.U.R. (Rossum's Universal Robots)* in which mechanical men created to do humanity's bidding take over the world and destroy their creators. This Frankenstein-like view of robots (or *robota*, meaning "forced work") was to become the prevailing view of robots in the fearful popular imagination.

It wasn't until the publication in 1939 of the story "I, Robot" by American science fiction author Otto Binder (1911–1974; writing as Eando Binder) that the world was introduced to Adam Link, an intelligent robot endowed with human emotions. "I, Robot" and its many sequels were the first robot tales to imagine "man-machines" that were not only smart, but also humanlike and nonthreatening in their behavior. Binder's stories inspired fellow science fiction author

Isaac Asimov's (1920–1992) own intelligent/emotional robot stories beginning in 1941 and collected in a single volume in 1950 under the title *I, Robot*. (This title was chosen by Asimov's publisher; the Adam Link stories were also collected and published in book form in 1965 under the title *Adam Link, Robot*.)

Binder's and Asimov's stories were captivating and inspiring to a whole generation of young readers. Many of them would translate their interest in science fiction to careers in science and technology. To them, the idea of friendly, helpful robots such as those popularized in the films and television programs of the 1950s to the present, including *Forbidden Planet, Lost In Space, Star Wars*, and others, were not a fantasy but a prediction of things to come.

Robots have become more common than most people think. That automated vacuum cleaner that operates by itself is a robot. So are those mechanical miniature cats, dogs, and humans sold as toys. Many industries—like the automotive industry—employ robots in the manufacture and distribution of goods. Robots also serve as stand-ins for humans in such dangerous activities as flying in surveillance aircraft, space and deep-sea exploration, handling explosives for police bomb squads, and mining. Indeed, robots are all around us.

To the new breed of robotic engineers, robots are both practical tools for home and work, as well as cutting-edge technology. Corporations, including the Honda automobile company, are hard at work pushing that technological edge with devices such as the humanoid robot Asimo (named in honor of author Isaac Asimov). Asimo can walk and run on two legs, observe and react to its environment, and learn to perform simple tasks independent of human control.

It will no doubt be the student participants in contests like the FIRST Robotics Competition who will take these remarkable robotic devices to the next level. With the proper education and training, today's young engineers will one day find themselves on the creative edge of a revolution in robotics and artificial intelligence, or AI.

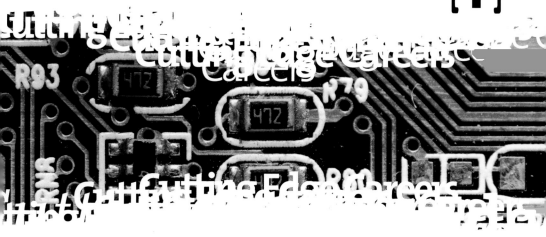

CHAPTER 1

The Range of Robotics Careers

Science remains decades away from the creation of robots that look, act, and think like human beings, but the goal of robotics has never really been to replicate humans. Rather, the goal is for robots to assist humans in their endeavors.

Robotics has evolved into a popular hobby that has spawned shelves of how-to books and aisles of home kits that allow anyone to build robots—ranging from simple, single-function devices to more complex machines capable of performing a variety of tasks. Dozens of competitions are held every year that allow students to demonstrate their skills as robotics engineers by designing and building working robots. Prizes, including scholarships, are often awarded to the winning amateur engineers.

A strong indication of the popularity robotics has achieved was the Comedy Central cable network program *BattleBots* (2000–2002), in which teams of competitors constructed fighting robots out of a junk pile of parts and sent them into battle to determine which team had created the superior machine. A BattleBots IQ National Competition, started by the producers of the TV program in 2001, is still held annually at Universal Studios in Orlando, Florida. Judges include high-ranking government education officials and NASA and military engineers.

"The student robot builders of today are the innovators of tomorrow," BattleBots IQ CEO Nola Garcia said about the entrants in the 2005 competition, as quoted in an article on DanceWithShadows.com. "You can see inventing better robots as a metaphor for inventing better lives; these students definitely exhibit the skills and attitudes that they need for future success, and will be leaders in the fields of science and engineering."

Whether your interest in robotics is sparked by science fiction or hands-on experience, this area of technology demands individuals who are both smart and creative. It requires not only knowledge of math, science, engineering, and the computer sciences, but also the imagination to see beyond that which has already been achieved in order to make the imaginative leap to the next technological level.

The Field of Robotics

The variety of jobs available in the broader field of robotics is extensive, covering everything from mathematics, electrical and electronics engineering, physics, systems and controls, computers, artificial intelligence (AI), computer-aided design and manufacturing (CAD and CAM), industrial engineering, and more. Careers in these areas involving both research and practical applications can be found in the aerospace industry, manufacturing, industrial production,

Robots have many practical applications in everyday life, from assembly lines to your living room floor. iRobot's Roomba Intelligent Floorvac is programmed to independently navigate and vacuum a room without human control or oversight.

electronics, the computer sciences, and, of course, robotics. Jobs exist in private industry, the military or government, and at the university level.

"Robotics engineer" is a title that was most often given to a manufacturing, mechanical, or electronics engineer who specialized in robotics and automated engineering systems. But as more and more people graduate from specialized studies in robotics, the title has become more closely associated with engineers whose focus is on the design of new robots and automated robotic systems, as well as those who research, design, and develop the next waves of robotic applications.

Robotics is one of the fastest-growing industrial fields. According to the U.S. Department of Labor, "Employers will need more engineers as they increase investment in plant and equipment to further increase productivity and expand output of goods and services" (as quoted in the "Engineering Occupations in Robotics and Automated Systems" section of California Occupational Guide Number 2004-A). This is a result of U.S. industries turning to cost-saving technologies—such as robots and automated systems—to improve their productivity and competitiveness with cheaper overseas labor. These developments have changed the way goods are produced and the skill requirements for factory workers.

The United Nations World Robotics Survey for 2006 predicts an optimistic future for both industrial and domestic robots. According to a 2004 Associated Press report, "The use of robots around the home to mow lawns, vacuum floors, and manage other chores is set to surge sevenfold by 2007 as more consumers snap up smart machines. That boom coincides with record orders for industrial robots . . . By the end of 2007, some 4.1 million domestic robots will likely be in use. Lawnmowers will still make up the majority, but sales of window-washing and pool-cleaning robots are also set to take off . . . [B]y 2007, world industrial robot numbers will likely reach at least 1 million."

Engineers

Engineers apply the theories and principles of science and mathematics to solve technical problems. They provide the link between scientific discoveries—for instance, the discovery of new information storage technology—and their practical applications in everyday use—integrating that technology into existing and new systems.

Engineers design machinery, products, systems, and the processes for manufacturing them safely and efficiently. They also design the computers that control automated systems for manufacturing, businesses, and the home. Engineers are also involved in more specialized areas of design, like robotic aids for the handicapped and the robotic equipment used in outer space, on the ocean floor, for military defense, and in other nontraditional areas.

The most common specialties in robot engineering include:

Mechanical engineers: Mechanical engineers are involved with the design, manufacture, and operation of robots and automated devices. These engineers must understand all aspects of mechanical functions, from gears to motors and the hydraulic or pneumatic systems that power and move many types of robots. (Hydraulic systems use fluids in motion to generate power and movement, while pneumatic systems rely on compressed air.)

Electrical and electronics engineers: Electrical and electronics engineers are the experts in the electrical systems that power and operate robots. They design, build, and test robotic systems. They may also specialize in automation controls, laser and optical systems, sensors, power systems, or electromagnetic fields and systems.

Industrial engineers: Industrial engineers specialize in designing and building factories and industrial systems. They are experts in productivity, determining how an industrial space can be built and used in the most efficient, safe, and cost-effective way possible.

CAD/CAM engineers: CAD/CAM engineers are the experts in computer-aided design and manufacture. They oversee the computers, robots, and automated systems used in the automated industrial process. Their systems run the assembly and production lines that manufacture products, such as cars, candy, toys, and electronics.

Manufacturing engineers: Manufacturing engineers work with CAD/CAM engineers, managing the machines and the production process in a manufacturing plant.

Computer engineers: Computer engineers design the architecture and hardware of the computer systems that control robots, analyze data processing requirements, and oversee hardware and software requirements.

Technicians and Specialists

Engineers come up with ideas and designs, but it is the technicians who get their hands dirty solving the problems of making those ideas work in the real world. According to the Center for Occupational Research and Development, technicians can be responsible for such tasks as installing and setting up robots, automated systems, computer systems, and programmable controllers. They measure robot performance and accuracy, as well as repair and test robots and automated manufacturing systems. The different specialties a technician can study include robot mechanics, electronics, production, manufacturing, machinery, engineering, and electromechanical and automation repair.

Technicians are needed both where robots are built and where they perform their programmed functions. While most technicians are employed by large manufacturers such as automakers and electronics and consumer goods producers, others are needed in the space program, law enforcement, and the service industry.

Like engineers, technicians should have a background in math and the sciences. In their youth, they will probably have shown an interest in mechanical and robotic toys and models. Technicians must be able to handle work tools and analytical instruments. They must have the patience and persistence to solve difficult problems and provide creative solutions to technical challenges.

Robots in Nissan's factory in Yokohama, Japan, assemble automobile engines and axles. Japanese industry uses 320 robots for every 10,000 employees, the highest ratio of automation in the world.

A crucial contributor to the creation of a working robot is the computer specialist. Computers are at the heart of every robot, carrying the programming and instructions that animate it and make it perform its tasks. Specialists design, program, and implement the computer systems, seeking ways to make computers smarter and better able to act independently of human oversight. They will also work on the problem of artificial machine intelligence, finding ways for computers to mimic human emotional responses.

But before anyone starts building robots that can explore the ocean floor or navigate their own way to Mars, years of training and education must first be acquired.

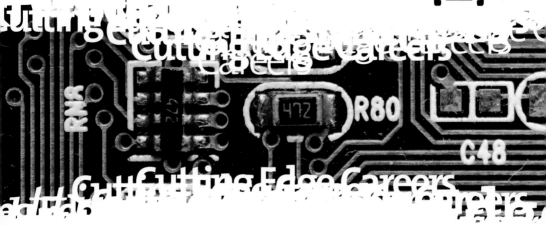

CHAPTER 2

Education and Training

Every engineer must have a solid education and vocational training in mechanics, electronics, hydraulics, pneumatic systems, computers, and other disciplines. The U.S. Department of Labor recommends a bachelor's or master's degree in engineering, with course work emphasizing robotics and automated systems. Advanced degrees are recommended for anyone planning to work in research and development, systems designs, or teaching and management positions.

High School

It's never too early to start planning for your career. Most high schools do not offer courses specifically designed for those

Events such as the Robotics Competition, which pits remote-controlled robots created by high school students against each other, offer future engineers the opportunity to not only practice their craft, but also to win college scholarships and earn impressive credits for college applications.

interested in robotics, but it is possible to pull together a credible course of study from the available classes. Algebra, geometry, trigonometry, and calculus are taught in every high school. Many schools offer advanced math courses for advanced college credits. In addition, local community colleges often allow high school students to enroll in higher-level courses.

Take as many science courses as you can, from general science and physics to biology and chemistry. If your school offers vocational courses, like mechanics, electronics, electrical engineering, computer programming, metal and wood shop, take as many as you can. Each one will teach you something you will need to know as a robotics technician and will demonstrate your commitment to becoming a well-rounded engineer.

That commitment can also extend to extracurricular activities such as hobbyist clubs. There are more than a million robot hobbyists in America who build their own robots and systems; millions more are interested in computers, electronics, and mathematical theory. Seek out these groups—either through your school, local youth organizations, or the Internet—and become involved. Clubs such as these provide the young engineer with hands-on experience and a competitive advantage when applying to colleges.

There are books available on every aspect of robotics imaginable in every school and public library, as well as magazines dedicated to electronics, computers, and robotics. Read as much as you can on your own—beyond what's required for school—on your particular areas of interest.

If, with all this studying, memberships in clubs, and extra reading, you still have time for a job, look for one related to electronics, computers, or mechanics that will allow you to use and expand your knowledge and skills. Internships offer academic credits for students who work at entry-level positions in private industry or government agencies. Internships place you in an actual work environment, side-by-side with experienced engineers, technicians, and managers.

Vocational and Technical Schools

Not all careers in robotics require four-year undergraduate or even more advanced degrees. There are many excellent technical or vocational schools that offer two-year programs in computers, mechanics, electronics, CAD/CAM, electromechanics, and related fields. These private trade and technical schools offer a variety of robotics-related courses, including engineering, automation, electromechanical technology, biomedical equipment repair, telecommunications, machinist training, electrical engineering, digital electronics, industrial computer technology, and industrial electronics.

Community colleges, trade schools, and technical schools should be fully accredited, which means they must follow an approved curriculum and be recognized by state and private educational agencies such as the National Association of Trade and Technical Schools (NATTS). To determine which school best suits your needs, you should request catalogs from those that interest you. Catalogs and curriculum might also be available online or in your guidance counselor's office. Your goal is to finish a trade or technical school with all the skills necessary to immediately begin a job in your chosen field. You want a school that teaches all the latest technology in modern facilities with up-to-date workshops and laboratories.

Tech and trade schools do not need to be the end of the educational line, however. What they do provide is a faster track to a career in your field, after which you can continue your studies part-time to earn advanced degrees. Community colleges and some trade schools offer two-year associate degrees, which are about halfway to a bachelor's degree. The next step would be a master's, which usually takes another one to two years to complete. Each rung up the educational ladder translates into an increase in career opportunities. Two-year degrees are available in most of the technical fields, including computers, electromechanical programs, manufacturing engineering, and biomedical engineering.

Higher Education

Advanced jobs in robotics require advanced degrees. You must start with a bachelor of science (B.S.) in engineering, computers, electronics, or another related field. You would then get a master's, and perhaps even a Ph.D. (doctorate) and postdoctoral work in an area of specialty. It bears repeating that it is never too soon to start studying for a career

The Massachusetts Institute of Technology (MIT) Artificial Intelligence Laboratory offers engineering students hands-on experience in cutting-edge robotics technology. The robot Kismet was operated by MIT from 1993 to 2000 and is now on display in the school's science museum in Cambridge, Massachusetts.

in high-tech fields. Begin in high school if you can by taking advanced and college preparatory courses in shop and laboratory practices, drafting, general science, computers, electronics, and mathematics. The competition for available spaces in engineering schools is strong, so proof of advanced course work on your transcript is important, as are extracurricular activities and good scores on the SAT or ACT.

There are many factors involved in choosing the college or university you will attend. Some are social, some are financial, and some are academic. But a college's reputation in the academic and corporate

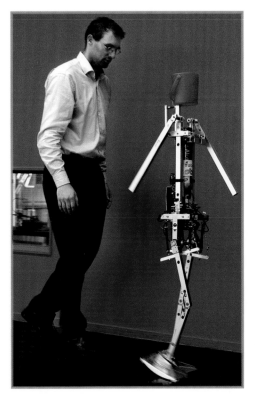

Designing robots that can mimic human motion was long a major engineering challenge. Denise, developed by the engineers at Delft University of Technology in the Netherlands, is one of the new generation of efficient humanoid robots.

worlds is one of the most important considerations. While not every school is as well-known in engineering circles as MIT, Stanford, Texas A&M, Cal Tech, Carnegie Mellon, or the University of Illinois, there are still many well-respected (and more affordable) schools offering excellent programs in engineering, robotics, and computers.

The Society of Manufacturing Engineers (SME) and similar engineering and technical organizations can also be helpful in identifying colleges and universities offering robotics and engineering programs. The school of your choice should be accredited by both a regional accrediting agency and by the Accreditation Board for Engineering and Technology (ABET). ABET is recognized as the primary agency for evaluating and accepting engineering and computer programs in American colleges and universities.

You should set your educational goals to the professional level to which you aspire. A B.S. will usually suffice in the commercialized fields of product design, manufacturing, marketing, and technical services. Research and development—the cutting edge of scientific development—usually requires at least an M.S. or Ph.D., as do most university teaching and research positions.

Guides to graduate programs, like *Peterson's Graduate Programs in Engineering and Applied Sciences*, are helpful in identifying the postgraduate program best for you. You will also be able to gather recommendations from instructors, professors, and professional colleagues and learn from their experiences about the best advanced degree programs available to you.

Scholarships and Grants

A college degree and postgraduate study are extremely expensive undertakings that require many students to seek out scholarships and other forms of financial aid to help reduce or defer the cost of their educations.

Scholarships and cash grants and awards are available from a wide variety of sources—from the government and the colleges themselves to private and public educational institutions, professional associations and corporations, and even contests. The FIRST Robotics Competition is one example of scores of contests that offer scholarships as prizes, as do science fairs and competitions sponsored by high-tech companies such as Intel, Microsoft, and GE. Many of these competitions are geared specifically toward robotics and engineering and offer the winner not only financial rewards, but also another item on his or her résumé to help influence a college admissions board.

The Society of Mechanical Engineers is one example of a professional organization that offers scholarships and grants and also has created its own education foundation "as a means of transforming manufacturing education in North American colleges and universities," according to the foundation's Web site. Since it began in 1980, the foundation has awarded more than $15.5 million in cash grants, scholarships, and awards for manufacturing education. Information on how to apply for these scholarships is available by contacting the granting organizations directly, many of which can be found in books such as *The College Board Scholarship Handbook 2007*, published by the College Board.

CHAPTER 3

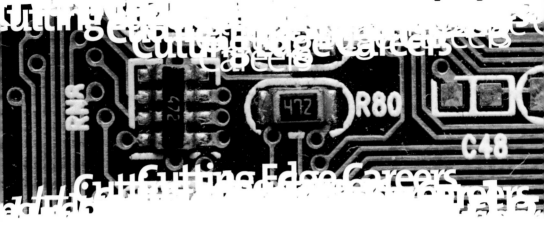

Desirable Skills, Qualifications, and Experience

Because robotics involves disciplines ranging from computer programming to electromechanics and industrial engineering, the future robotics engineer has to possess a broad range of knowledge. An engineer cannot design a mechanical robotic device without a full understanding of how the mechanical parts interact with the electronic and hydraulic parts that are controlled by the computer. Nor can a technician be expected to repair a device if he or she doesn't know how all the pieces fit together and operate.

What makes a good engineer? To start, a good engineer has varied tools and a keen attention to detail. Future engineers are those who take an early interest in how things work and are fascinated by electronic and mechanical devices. Engineering

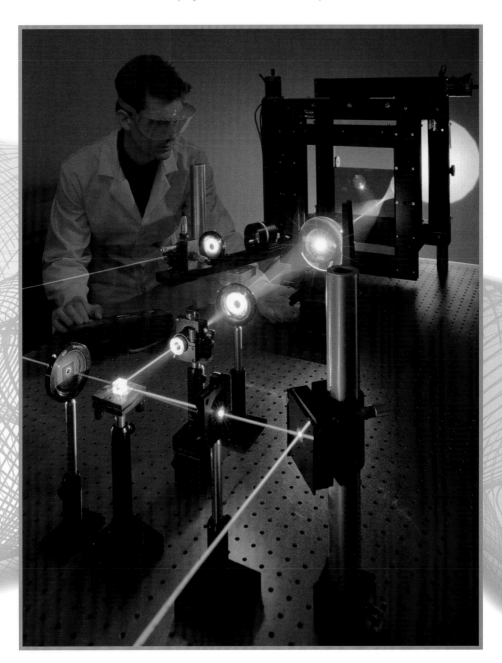

Robots capable of operating independently of direct human control must be able to "see" their environment. The study of lasers and optics is a vital component of this goal, enabling robots to recognize and navigate around obstacles.

prospects generally do well in math and science courses and score above average in those subjects on the SATs or ACTs.

Their hobbies might include anything from tinkering with mechanical devices, robotic kits, and model rockets in the garage and backyard to science fiction, computer programming or 3-D modeling, and role-playing fantasy games. All these pursuits build an active imagination and help develop the qualities an engineer should possess, according to the U.S. Department of Labor. These include analytical thinking and developed research skills, computer literacy, creativity, patience, the persistence to solve problems logically, a capacity for detail and accuracy, and good communication and interpersonal skills.

Employers' Expectations and Requirements

When you interview for a first job, employers are not just looking at your grades, test scores, and awards. They are looking at *you*, to see if you are a well-rounded individual who can bring more than just basic skills to the table. That means your knowledge and experience aren't rooted solely in engineering, computers, electronics, or whatever your chosen specialty happens to be. An engineer must be able to write a competent English composition and have taken courses in drafting, mathematics, the basic sciences, physics, and a dozen other disciplines and fields.

A potential employer will be primarily interested in seeing that you have acquired adequate academic training and have taken all the courses needed to work in the field of robotics. The Accreditation Board for Engineering and Technology provides a recommended pre-engineering program for the freshman and sophomore years of college that includes freshman chemistry, general physics, English, mathematics (including calculus, analytic geometry, and differential equations), engineering graphics, applied mechanics, and electives

Competitors in the 2005 National Science Olympiad at New Trier High School in Winnetka, Illinois, design and build robots capable of simple tasks such as locating and picking up golf and Ping-Pong balls from a playing field.

from the social sciences, humanities, matrix algebra, computers, materials, and statistics.

Once accepted into an engineering program, the curriculum consists of courses in analytical mechanics and engineering analysis, calculus, physics, principles of electricity and magnetism, quantum theory and solid state physics, materials science, chemistry, engineering science (which includes drafting and descriptive geometry), and electrical engineering. Along with this course load are such additional required courses as hydraulics and pneumatics, manufacturing management, industrial design, computer controllers and architecture, laser and microwave theory, digital electronics and design, statistics, mechanical engineering, computer programming, and artificial intelligence.

And to top that all off, an employer wants to see a prospective employee involved in extracurricular—and not necessarily engineering-related—activities. Clubs and groups, competitions, and social and charitable activities are also important and show a boss-to-be that this applicant is a dedicated, conscientious, energetic, socially engaged worker.

Most colleges and postgraduate schools provide placement assistance for new graduates. Job placement specialists in the high-tech industry are always in search of promising candidates for a wide variety of robotics-related positions. Professional organizations, online job search sites, internships, and personal contacts and networking are all excellent sources for jobs. Candidates can also contact a company or research facility that creates products or services they admire, provide their résumé, and seek an informational interview.

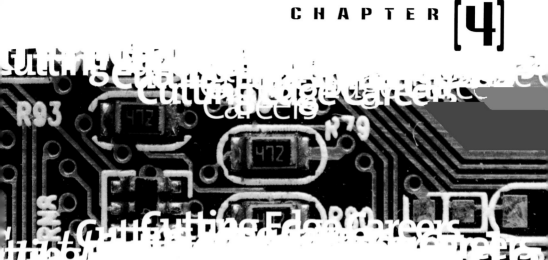

CHAPTER 4

Robots at Work

Robots that once existed only in the fertile imaginations of science fiction writers are today becoming more and more a part of everyday life. From assembling cars on the factory floor to vacuuming our living room carpets, the demand for robots is on the rise. So, too, are the opportunities for the robotics engineers who create and build them.

According to the U.S. Bureau of Labor Statistics and the U.S. Department of Labor, for each year from 2002 through 2012, employers in all manufacturing industries are projected to need 17,000 industrial and manufacturing engineers, 2,000 materials engineers, 14,000 mechanical engineers, 7,000 industrial engineering technicians, 7,000 mechanical engineering technicians, and 273,000 metal and plastics workers.

Robots can replace humans in hazardous situations. The Mark V is an 800-pound robot that can be operated by remote control up to one-half mile away. In this demonstration, the Mark V examines a suspicious briefcase that may contain a bomb.

This includes computer programmers and operators, machine operators and tenders, machinists, and welding and soldering workers.

Japanese industry currently uses 320 robots per 10,000 employees, while Germany uses 148, Italy 116, Sweden 99, and between 50 and 80 each in the United States, Finland, France, Spain, Austria, Denmark, Belgium, the Netherlands, and Luxembourg. In order to remain competitive in the global market, American companies will need a growing number of robotics engineers to help automate their factories.

According to a 2004 report by the California Employment Development Department, the development and use of robots are spreading beyond factory floors where they first appeared several decades ago. "The 'service' or mobile robot industry is growing and these new applications and innovations demand new skills," the report noted. The number-one skill in demand is that of robotics engineer. The engineer is responsible for designing industrial robots and automated systems, as well as specialized robots in a number of fields.

Jobs for industrial robotics engineers can be found in the automobile industry, the health sciences, aerospace, agriculture, bioengineering, chemicals, computers, electronics or electrical, engineering physics and mechanics, heavy industry, food manufacturing, mining, oil processing, maintenance services, and remote exploration. In addition to assembling cars and computers, robots are being used to perform tasks ranging from the packaging, labeling, and wrapping of products to the moving and storing or disposing of toxic chemicals and other hazardous materials. The military uses robots for surveillance, missile navigation, and the handling of unexploded bombs.

What follows is a sampling of the many uses for robots and the specialized tasks for which they are being designed. Each project described represents a wide array of intriguing jobs for someone trained in robotics engineering.

Planetary Rovers and Space Probes

Robots have already been sent to places too dangerous or difficult for humans to reach, including deep into mines, the ocean floor, and, most famously, outer space. The Mars Rover that was launched by NASA in 2003 and began exploring the Martian surface in 2005 is a robot geologist. It collects rock and soil samples and studies the planet's surface by camera. It uses microscopic imagers for close-up

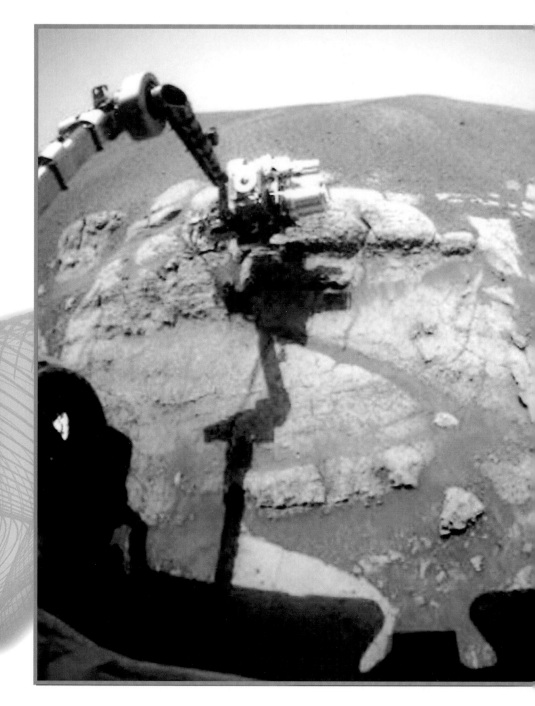

Another example of robots replacing humans in dangerous places is in space exploration. In 2004, NASA's Mars Exploration Rover Opportunity explored the Martian surface and employed a variety of tools to examine rock samples in its search for signs of life on the red planet.

NASA scientists use a Mars Exploration Rover (MER) in their Pasadena, California, laboratory, which mimics the surface conditions of Mars to plan and test movements for the two rovers operating on the Martian surface.

views of the soil and spectrometers to analyze the soil's mineral content. The rover utilizes a robotic arm and a magnetic array to gather its samples, all independent of human control.

The rover was not NASA's first experience with robotic components. NASA has launched robotic space probes, including *Cassini* to Saturn. Before beginning its orbit around Saturn, *Cassini* launched the Huygen robot probe to Titan, one of Saturn's moons, in 2004. The *Stardust* robotic probe intercepted the comet Wild-2 in 2005 and, after a three billion mile (4.8 billion kilometer) round-trip and a collision with the comet, returned two minutes ahead of schedule to its landing site southwest of Salt Lake City,

Utah. The space shuttle's robot arm, used to both launch and retrieve satellites and cargo from the shuttle, was also an early NASA success in robotics.

RoboGuards

A German company, Robowatch Technologies, has developed MOSRO (Mobile Sicherheitsroboter), a robot security guard. MOSRO patrols indoor areas such as malls, parking garages, and factories. It can be equipped with up to 240 sensors for detecting motion, smoke, heat, and gas. It can issue verbal warnings to troublemakers in more than twenty languages. OFRO (Mobile Freilandüberwachung) is MOSRO's partner, working outdoors to patrol the perimeters of military bases, airports, industrial complexes, and other large areas. OFRO is weatherproof and can detect intruders with multiple sensors and thermo-cameras.

Exploration and Innovation

Both academia and industry have developed robots for a wide variety of research and product innovation. In 1994, Carnegie Mellon University created Dante II, an eight-legged robot that was used to go into an active volcano—Mount Spurr in Alaska's Aleutian Mountain Range—and collect samples. Submersible robotic craft, capable of autonomous (self-directed) movement, are used for deep-sea exploration. One such craft was used to locate the H.M.S. *Titanic*, the ocean liner that sank in the North Atlantic off Newfoundland in 1912 leading to the deaths of more than 1,700 passengers. Shell Oil and Textron developed a fully automated gas pump that opens your car's gas tank, fills it, and sends you on your way without any human assistance or involvement. The pumps have been tested in Westfield,

Robotics will one day help those suffering from paralysis or mobility problems. The Hybrid Assistive Leg (HAL) is a robotic exoskeleton developed at Japan's University of Tsukuba. The suit has sensors that read wearers' brain signals being sent to their muscles and anticipate and assist their movements.

Indiana, and Sacramento, California, but have not yet been implemented for use nationwide.

Surgery and Prosthetics

Robotic and computer technology is now so advanced that robots are even being used to perform surgery, their movements guided by a human surgeon via computer. The robot surgeon scales down the human surgeon's movements, allowing the doctor to perform delicate surgeries in inaccessible areas of the body and using movements far too fine and precise for a human hand to execute.

Engineers at Yale University have developed a similar robotic tool, known as the da Vinci system. Its multiple arms are outfitted with microtools for precise work. Robotics engineers are also involved in developing the next generation of prosthetics. Robotic hands, arms, and legs will be developed for human use, thanks to the work being done on humanoid robots that are able to mimic human movements. These artificial appendages will allow those who have lost a limb from illness, accident, or birth defect to be fitted for a prosthetic that allows for near normal movement.

Service Robots

Currently, robots are being used to clean sewers, inspect the aluminum skin of jet airliners, mow lawns, and vacuum homes and industrial sites. These domestic robots are equipped with sensors that allow them to avoid collisions and maneuver without incident. Wakamaru, by Mitsubishi, can recognize faces and speech and is designed to care for the elderly and disabled. Robots of all kinds are being marketed in toy and hobby stores: robot pets, butlers, gladiators, and vehicles.

Engineering a Robot's Thoughts and Senses

To accomplish all these tasks and functions, robots must be able to move, maneuver, see, hear, and analyze—or "think through"—their tasks. The goal in industrial, consumer, and scientific applications of robots is to create machines that imitate human movements. But every task—whether large-scale and dependent on brute strength, or small-scale and delicate—is the result of a complicated series of computer commands, electronic impulses, and mechanical responses.

The kind of precision work that gets a specific part onto the assembly line exactly when and where it is needed increasingly requires more and more sophisticated optical and computer systems. These systems must handle the increased flow of information and assist the robot in processing and making use of it. Robots need to hear as well as see; they are even beginning to speak. They interpret spoken words and commands with sophisticated voice recognition programs and respond by performing the requested tasks. Sometimes they respond with preprogrammed statements in voices that are becoming increasingly difficult to distinguish from those of humans.

For most robots, "thinking" involves a fairly simple and repetitive task, like automated welding on an auto production line or moving merchandise around a warehouse. Increasingly, however, there is a need for smarter robots, outfitted with enough processing power to perform more complex functions.

Engineers use the example of tying a shoelace to explain the enormous complexity of teaching a machine to perform even relatively simple actions. Imagine you are explaining to a friend, who is blindfolded and holding a pair of pliers in each hand, how to tie his shoelace. Since he's blindfolded, you can't simply tell him to pick up the ends of the shoelace with the pliers. You first have to instruct him in every single, simple motion required to locate the laces by exact coordinates on a three-dimensional grid, then each motion required to open the pliers and where along the length of each lace to pick up the lace, how high to lift each end, the degree of the arc that your friend's one hand has to make on the three-dimensional grid in order to cross the laces, then the distance along the length of each end of the lace where they should intersect—the process would be endless.

But that's precisely what robotics engineers are attempting to do—not just for a single movement or task, but also for all the tasks performed, consciously or otherwise, by humans. These can range from grasping a hammer or holding an egg without breaking it to walking up a flight of stairs or running through a crowded space without collisions. Many large companies are investing billions of dollars in robotics research, some attempting to create lifelike robots like Honda's Asimo, Fujitsu's HOAP-1, and Sony's QRIO. Each of these robots can, among other things, walk upright on two legs, mimic the human range of motion, navigate autonomously, and avoid obstacles.

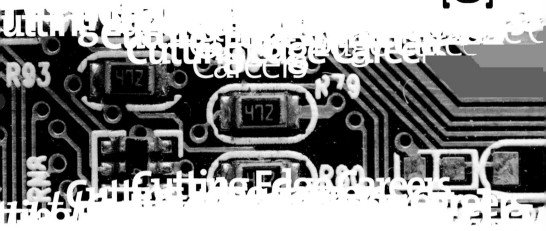

CHAPTER 5

Building a Working Robot Step-by-Step

A robot is like a jigsaw puzzle with a million pieces, all of which must fit smoothly together if you wish to see the entire picture. The more complex a robot is, the more technology it requires to perform its functions.

As mentioned in the previous chapter, some companies are investing in the research and development of humanlike robots. The uses for a fully functioning, autonomous robot capable of learning its tasks through instruction and repetition are almost infinite. Such "smart" robots can be employed in hazardous industrial positions, in exploration of inhospitable terrain or environments (such as outer space or the ocean floor), or in numerous military applications (such as bomb defusing or mine sweeping). They can also be employed as household

Japan's Honda Motor Company's humanoid robot, Asimo, can not only walk like a human, but can also maintain its balance while kicking a soccer ball. Dozens of engineering and scientific disciplines went into the creation of Asimo.

servants, performing menial tasks that their owners don't wish to do themselves.

One of the more advanced examples of a humanoid robot is the Honda Motor Company's Advanced Step in Innovative Mobility, or Asimo. Almost every aspect of robotics engineering has been employed in the development of this one machine. In learning how a fully functioning, sophisticated robot is designed, developed, and tested, you will gain some insight into the range of jobs, specializations, and tasks that are available to robotics engineers.

Engineering and Mechanics: Teaching a Robot to Walk

Asimo began in 1986 as a technological and design challenge put to the engineers of the Honda Motor Company. Seeking to develop revolutionary new technology, the engineers first had to decide on the ideal tasks for a walking robot to perform in human society. Their goal was to create a robot helper for hazardous or menial jobs, for housekeeping chores, or to assist in the care of the elderly and handicapped. The robot would have to be able to walk up and down stairs, maneuver on uneven terrain and around objects in a crowded room, and access the same places that humans do. Ultimately, it was determined that the best model for the robot's design would be the human body itself.

One of the engineers' greatest challenges was duplicating the walking motions of humans. We don't have to think consciously about the endless series of movements that allow us to walk upright—maintaining our balance, shifting our weight just so to compensate for each step or turn, recovering our balance after a misstep. The process has become reflexive, but watch a baby just learning to walk and you will quickly see how complicated and difficult a process it can be.

Before they could build their robot, Honda's engineers first had to see if they could make it walk. They studied a whole range of motion, from the legs of insects to the motion of a mountain climber with prosthetic legs (which more closely resemble robotic legs). From these, they learned all of the many, often almost imperceptible movements that are made—from our heads down to our toes—when we walk. One of the things they learned during this study was that we shift our weight with our entire bodies to maintain balance, and that the toes of the human foot are helpful in this process. They then designed a set of robotic legs to put what they had learned into practice, trying to duplicate human motions through mechanical means. They even gave Asimo soft projections on its feet that act as its toes.

The Honda engineers studied the way joints function in the human body and incorporated them into Asimo in the form of servo-motors—small but powerful motors with a rotating shaft that moves limbs or surfaces to a specific angle as directed by a controller. Once the motor has turned to the appropriate angle, it shuts off until it is instructed to turn on again. Asimo includes twenty-six joints, or "degrees of freedom," as engineers call them. A single degree of freedom allows for movement either right and left or up and down. Asimo has two degrees of freedom in its neck and six in each arm and leg.

Asimo went through several early incarnations, developing from a pair of legs that would take as much as twenty seconds to take a single shuffling, or "static," step to the much more humanlike "dynamic" walking. Dynamic walking meant that Asimo leaned into the next step, shifting its weight and moving the other foot forward to catch itself, so that rather than falling forward, it walked forward. By the sixth prototype, the engineers had solved the problem of walking and had a robot that could walk up an incline, up stairs, and on uneven terrain. With the most recent Asimo Walking Technology, the robot can even predict its next movement in real time and shift its center of gravity in anticipation of it.

Once they had a machine that could walk, the engineers needed an upper body and head for their creation. Once again, they chose the human body as the basic model for Asimo, initially making him six feet, two inches (188 centimeters) tall and 386 pounds (175 kilograms). Successive models were scaled down to four feet (122 cm) and 115 pounds (52 kg). At that size, the engineers reasoned that Asimo would be less intimidating to people, even those who are seated, but still tall enough to work at a standard table or counter and reach doorknobs and light switches.

Computers

All the revolutionary engineering in the world is useless without the computer power to instruct all the parts when and how to move. Asimo is equipped with a speed sensor and a gyroscope sensor that continually monitor and correct its body position and the speed at which it is moving. These instruments relay any information on necessary adjustments to the central computer. The computer then instructs Asimo's mechanical structure to make the proper corrective movements to maintain its balance.

Controllers and Artificial Intelligence

For all its sophistication, Asimo is not an autonomous robot. It cannot make its own decisions but instead relies on preprogrammed routines, human manual control, or voice commands to perform its functions. Using wireless technology, an operator can control Asimo by a computer, even seeing what Asimo sees through its camera eyes. A joystick moves the robot like a radio-controlled car, and the operator

The applications for an independently acting robot are limitless. Asimo's descendants may one day be routinely involved in household chores, child care, the care of the sick and elderly, routine office work, exploration, and hazardous occupations such as mining.

controls its direction. Should Asimo encounter an obstacle or incline, it will automatically adjust to the terrain. Asimo carries its own computer, located in its backpack, to provide the processing power necessary to keep this complex machine on its feet.

Asimo is Internet-ready and can be integrated with a computer network. While artificial intelligence is not currently a part of Asimo's development, one day Asimo or its descendants will be able to decide and act upon their own course of action, without any human supervision, prompts, or commands.

Materials

Earlier versions of Asimo weighed as much as 463 pounds (210 kg), but as the engineers improved on the technology and reduced the robot's height, they started experimenting with new, lightweight materials for Asimo's "bones" and "skin." Thanks to such weight-saving materials as its magnesium-alloy body and plastic skin, Asimo weighs in at a surprisingly light 115 pounds (52 kg).

Optics

Robots "see" in much the same way that humans do. Instead of eyes, however, they have cameras that capture the images and send them to the "brain"—the computer—by wires (similar to our optic nerves) where the information can be interpreted and reacted to. In the controlled environment of a manufacturing plant, where a robot might be programmed to visually inspect items on a conveyor belt or navigate a limited and predetermined route, the problem of vision requirements is not very complex.

However, for a robot expected to operate alongside people in the real world, it must have a wide and well-defined field of vision. Asimo must be able to navigate safely inside homes and buildings and on the street. It must be able to "understand" what it sees, distinguishing shadows from real objects, detecting and recognizing objects in real time by their size, shape, and color (comparing them to the thousands of items stored in its database). It must also detect multiple objects and calculate their distance, trajectory, and speed—a valuable skill to have when you wish simply to cross the street.

Environmental Recognition

Thanks to its ability to see and understand its environment, Asimo is surprisingly aware of its immediate surroundings and can react effectively to them. With its camera eyes, the robot can detect and track movement, recognize faces to greet friends, and interpret hand motions such as obeying the "stop" hand gesture. It can even recognize when it is being offered a hand to shake or when someone is waving at it.

Asimo can also interpret its surroundings and act in a manner that is safest for both it and nearby humans. It will recognize potential hazards, like stairs, and react accordingly to them. For instance, it will stop and start to avoid colliding with humans and other moving objects.

Sound and Voice Recognition

Asimo can distinguish between different voices and other sounds. It can respond to its name, face people who are speaking to it, and even react to sudden, loud noises by turning in the direction of the sound. Asimo understands voice commands as well, with a database of spoken commands that activate various preprogrammed movements.

Power Source

Without a power source, Asimo is just a complicated sculpture. Currently, its power source is a rechargeable, forty-volt, nickel metal

Asimo leads Japanese elementary school children in a game of Simon Says. By bringing the advanced humanoid robot into classrooms, Honda Motor Company's remarkable engineering feat may inspire the next generation of robotics engineers.

hydride battery that lasts for thirty minutes on a single charge. However, the battery takes four hours to fully charge, so additional batteries are necessary if Asimo is to operate for a long stretch. The weight of the battery, stored in Asimo's midsection, helps create Asimo's center of gravity, keeping it stable and balanced.

Cybernetics

In 2006, engineers and scientists from Honda's ATR Computational Neuroscience Laboratories used an MRI (magnetic resonance imaging) machine to control Asimo simply by thinking. They had developed an interface (a means of communication between humans and machine) that measures a person's brain signals. These signals are then relayed to Asimo. In the test, the person in the MRI first made a fist and then a V-symbol (a peace sign) with the index and middle fingers. Asimo imitated both gestures just seconds later. A similar system could eventually be used by humans to control all manner of machines, from a keyboard to bionic implants that could allow a person with spinal cord injuries to move their limbs simply by consciously thinking motion commands.

CHAPTER 6

The Future of Robots and Robotics

The future for robotics seems unlimited. While it's probably not reasonable to expect a world populated by humans and humanoid robots living and working side-by-side anytime soon, intelligent machines are nonetheless becoming increasingly common in all areas of our life and work.

In his 2002 book, *Flesh and Machines: How Robots Will Change Us*, Rodney A. Brooks, director of the MIT Artificial Intelligence Laboratory, predicted that within twenty years we will have robots that can think and experience emotions. Before we reach that advanced stage of development, robots will continue to become more autonomous and sophisticated. Colin Angle, the inventor of the robot vacuum cleaner Roomba, believes that inexpensive mechanical, self-directed devices

Nanobots, robots only one 100-billionth of a meter big, may one day revolutionize medicine. These miniscule machines could be programmed to attack specific cells, like cancer cells, and eradicate them without the use of harmful chemicals or radiation.

such as his vacuum and robotic toys will usher in an age of robotics without our even realizing it. In an article published by ExtremeTech.com entitled "Future Vision: Cheap Robots Change the World," Angle says, "We wake up one morning thinking about the past and realize that the things we take for granted are exceptionally different than they were when we were younger. In time, just as innovations like the light bulb and telephone elevated life as we know it to new standards, so will robotics."

The companies that solely manufacture and sell robots are growing in number. There are also many job opportunities with companies that manufacture peripheral robotics equipment, like parts and software, and in industries that use robotics. These industries include the automotive, aerospace, electronics, food processing, apparel manufacturers, and pharmaceutical industries.

According to California's Employment Development Department, almost half of all engineering jobs are located in manufacturing industries. As companies increase investment in their manufacturing plants and equipment to further boost productivity and increase output of goods and services, they will need more engineers. Additional career opportunities will develop in the service robot industry and in related technologies such as artificial intelligence, simulation, and machine vision. Robots will also continue to be made smarter, more reliable, and smaller. Nanobots—robots smaller than 100 nanometers (or 100 *billionth* of a meter)—are being developed with countless medical and industrial applications and promise to be a burgeoning technology in the twenty-first century.

"After a quarter-century of being involved with robotics, I have concluded that the robotics industry is here to stay," wrote Donald A. Vincent, executive vice president of the Robotics Industries Association (RIA), in the *Handbook of Industrial Robotics*. "Robotics, robots, and their peripheral equipment will respond well to the challenges of space construction, assembly, and communications; new applications

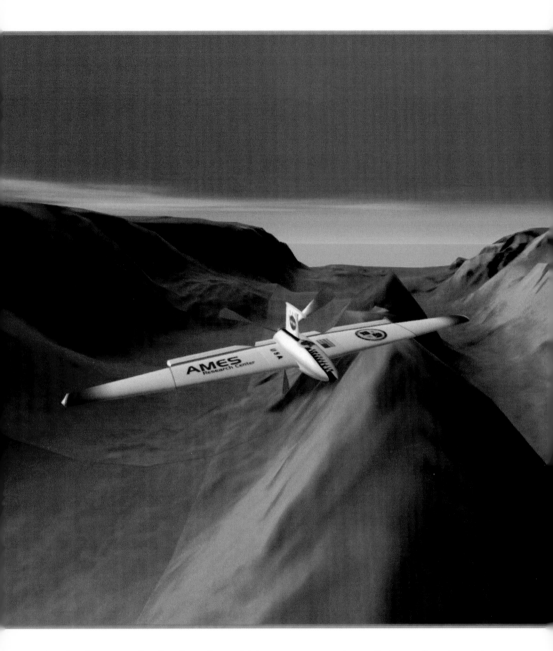

In the future, robotic aircraft could be sent to explore distant planets without the danger and expense of sending human astronauts. One such aerospace craft under development by NASA might one day soar through the canyons of Mars.

in agriculture, agri-industries, and chemical industries; work in recycling, cleaning, and hazardous waste disposal to protect our environment and the quality of our air and water; safe, reliable, and fast transportation relying on robotics in flight and on intelligent highways. Robotics prospered in the 1900s; it will thrive and proliferate in the twenty-first century."

And to meet that demand and make that proliferation possible, robotics engineers will be needed more than ever before. Like the discoverers of fire, the inventors of the wheel, and the creators of the first airplanes, radios, televisions, and computers, robotics engineers will be at the very center of a seismic shift in human perception and possibility. They will not only be at the cutting edge of one of the world's most stimulating and daring careers, but also will be at the forefront of a tremendous cultural shift that will change life as we know it and how it is lived. Now is the time to get involved in this dynamic and revolutionary field of endeavor.

GLOSSARY

artificial intelligence (AI) A branch of computer science that studies how to endow computers with capabilities of human intelligence.

automation The art of making processes or machines self-acting or self-moving.

automaton A mechanical device, usually powered by water, wind power, or clockworks.

computer An electronic device that stores, retrieves, and processes data and can be programmed with instructions.

computer-aided design (CAD) The use of a computer in industrial design applications such as architecture, engineering, and manufacturing.

computer-aided manufacturing (CAM) The use of computers to program, direct, and control production equipment in the fabrication of manufactured items.

cybernetics The science of control and communication in mechanical and biological systems.

electromechanics In engineering, electromechanics combines the sciences of electromagnetism, electrical engineering, and mechanics.

electronics The branch of physics that deals with the emission and effects of electrons and with the use of electronic devices.

engineer An engineer is someone who practices the engineering profession, using scientific knowledge to solve practical problems and produce goods for society.

gyroscope Gyroscopes are used extensively to sense angular movement and stabilize sensor platforms.

hydraulics Engineering science pertaining to liquid pressure and flow.

nanotechnology A branch of science and engineering devoted to the design and production of extremely small electronic devices and circuits built from individual atoms and molecules.

pneumatic Powered or inflated by compressed air.

robot A machine that can automatically do tasks normally controlled by humans and is mostly used to perform repetitive tasks on an assembly line. The term was first coined by Czech playwright Karel Capek in his 1921 play *R.U.R.* to describe independently acting mechanical devices.

simulacra A device created in the image of a living thing.

FOR MORE INFORMATION

FIRST Robotics Competition
200 Bedford Street
Manchester, NH 03101
(603) 666-3906
Web site: http://www.usfirst.org

Institute of Electrical and Electronics Engineers (IEEE)
Service Center
445 Hoes Lane
Piscataway, NJ 08854-4141
(732) 981-0060
Web site: http://www.ieee.org

Institute of Industrial Engineers
3577 Parkway Lane, Suite 200
Norcross, GA 30092
(770) 449-0460
Web site: http://www.iienet.org

Robotic Industries Association
900 Victors Way
P.O. Box 3724
Ann Arbor, MI 48106
(734) 994-6088
Web site: http://www.robotics.org

The Robotics Institute
5000 Forbes Avenue
Pittsburgh, PA 15213-3890
(412) 268-3818
Web site: http://www.ri.cmu.edu

Society of Manufacturing Engineers
One SME Drive
P.O. Box 930
Dearborn, MI 48121
(313) 271-1500
Web site: http://www.sme.org

Web Sites

Due to the changing nature of Internet links, Rosen Publishing has developed an online list of Web sites related to the subject of this book. This site is updated regularly. Please use this link to access the list:

http://www.rosenlinks.com/cec/robo

FOR FURTHER READING

Arrick, Roger. *Robot Building for Dummies*. New York, NY: For Dummies, 2003.

Branwyn, Gareth. *Absolute Beginner's Guide to Building Robots*. Indianapolis, IN: Que, 2003.

Bridgman, Roger. *Robot*. New York, NY: DK, 2004.

Brown, Jordan D. *Robo World: The Story of Robot Designer Cynthia Breazeal*. New York, NY: Franklin Watts, 2005.

Cook, David. *Robot Building for Beginners*. Berkeley, CA: Apress, 2002.

Domaine, Helena. *Robotics*. Minneapolis, MN: Lerner Publishing Group, 2005.

Eckold, David, ed. *Ultimate Robot Kit*. New York, NY: DK, 2001.

Gibilisco, Stan. *Concise Encyclopedia of Robotics*. New York, NY: McGraw-Hill/TAB Electronics, 2002.

Grand, Steve. *Growing Up with Lucy: How to Build an Android in Twenty Easy Steps*. London, England: Phoenix House, 2005.

Jones, David. *Mighty Robots: Mechanical Marvels That Fascinate and Frighten*. Toronto, ON: Annick Press, 2005.

McComb, Gordon. *Robot Builder's Sourcebook: Over 2,500 Sources for Robot Parts*. New York, NY: McGraw-Hill/TAB Electronics, 2002.

Predko, Myke. *123 Robotics Experiments for the Evil Genius*. New York, NY: McGraw-Hill/TAB Electronics, 2004.

Sobey, Ed. *How to Build Your Own Prize-Winning Robot*. Berkeley Heights, NJ: Enslow, 2002.

Stone, Brad. *The Turbulent Rise of Robotic Sports*. New York, NY: Simon & Schuster, 2003.

Williams, Karl. *Build Your Own Humanoid Robots: Six Amazingly Affordable Projects*. New York, NY: McGraw-Hill/TAB Electronics, 2004.

Williams, Karl. *Insectronics: Build Your Own Walking Robot*. New York, NY: McGraw-Hill/TAB Electronics, 2002.

Wise, Edwin. *Robotics Demystified*. New York, NY: McGraw-Hill Professional, 2004.

BIBLIOGRAPHY

Angle, Colin. "Future Vision: Cheap Robots Change the World."
ExtremeTech.com. September 17, 2002. Retrieved May 2006
(http://www.extremetech.com/article2/0,3973,538588,00.asp).

Associated Press. "UN Predicts Boom in Robot Labor."
CBSNews.com. October 20, 2004. Retrieved May 2006
(http://www.cbsnews.com/stories/2004/10/20/tech/
main650274.shtml).

Baker, Christopher W. *Robots Among Us: The Challenges and
Promises of Robotics*. Brookfield, CT: The Millbrook Press, 2002.

"Battlebots IQ 2005 Robot Competition Begins."
DanceWithShadows.com. April 13, 2005. Retrieved May 2006
(http://www.dancewithshadows.com/tech/
battlebots-iq-competition-2005.asp).

Brooks, Rodney A. *Flesh and Machines: How Robots Will Change Us*.
New York, NY: Pantheon Books, 2003.

"Careers in Robotics: Software Engineer." NASA. March 2003.
Retrieved May 2006 (http://robotics.nasa.gov/students/features/
bluefin.htm).

"Engineering Occupations in Robotics and Automated Systems."
California Employment Development Department. 2004.
Retrieved May 2006 (http://www.calmis.cahwnet.gov/file/
occguide-archive/engrobot.htm).

Fritz, Sandy. *Robotics and Artificial Intelligence*. North Mankato,
MN: Smart Apple Media, 2003.

Huse, Brian. "How Robots Will Affect Future Generations."
RoboticsOnline. Retrieved May 2006 (http://www.roboticsonline.
com/public/articles/details.cfm?id=600).

Malone, Robert. *Ultimate Robot*. New York, NY: Dorling Kindersley
Books, 2004.

Marrs, Texe W. *Careers with Robotics*. New York, NY: Facts on File, 1988.

Obringer, Lee Ann. "How ASIMO Works." HowStuffWorks.com.
Retrieved May 2006 (http://electronics.howstuffworks.com/
asimo.htm).

"Robots and TV to Be Big in 2006." BBC News. October 19, 2005.
Retrieved May 2006 (http://news.bbc.co.uk/1/hi/technology/
4357352.stm).

INDEX

About the Author

Paul Kupperberg is a New York-based writer who has written extensively on technology-related issues, including the space program, spy satellites, Edwin Hubble and the Big Bang, the science of disease, and the history of scientific innovation. He lives in Connecticut with his wife, Robin, and his son, Max.

Photo Credits

Designer: Evelyn Horovicz; Photo Researcher: Hillary Arnold